EG Practice
Flood Risk Assessment

Gareth W Evans

Estates Gazette, 1 Procter Street
London WC1V 6EU
A division of Reed Business Information

EG Books

Commissioning Editor: Alison Richards
Typesetting: Isabel Micallef — in Meta and Palatino
Interior design: Deborah Ford, EG
Cover design: Ted Masters
Printed and bound by: Short Run Press
Cover images: Gareth Evans

Contents

Acknowledgements, iv

Preface, v

1 Introduction
Definition of floods, 1
The significance of floods, 1

2 River and Stream Flooding
Basic hydrology, 5
What happens to rainwater, 5
The development of floodplains, 8
Different flood types, 11
Groundwater flooding, 11
Global warming, 11
Controlling and forecasting floods, 12
Perception of risk, 12

3 Marine Flooding
Sea defences, 14
Shrinkage, 15
Storm surges, 15
Sea level changes, 15
Perception of risk, 16

4 Unusual Flood Hazards
Dams, 17
Mine bursts, 17
Tsunami, 18

5 Case Studies
Lewes — 2000, 19
York — 2000, 21
Boscastle — 2004 (and Lynmouth — 1952), 22
Towyn — 1990, 23

6 Assessing Flood Risk for a Site
Client requirements, 25
Personal knowledge, 25
Research, 25
A sample Environment Agency map, 27
A simple exercise, 28
Reporting the risk, 28

Index, 29

Acknowledgements

Acknowledgements are particularly due to my wife Jan, who commented searchingly on my rough drafts and prepared the index; to Roger Edwards for kindly making his photographs available; and to David Murphy and Jayne Smith from the Environment Agency.

Preface

Every year seems to bring a flood disaster to the UK, be it large or small. Each time, I wonder why flooding is not better understood and some of the damage avoided. As this volume goes to press, the dreadful damage and loss of life inflicted on New Orleans and its neighbours serves grimly to underline the risk of the massive losses that can arise from an unfortunate combination of vulnerable landscape and extreme weather.

This publication is designed to provide basic information in an accessible package for surveyors and their professional relatives, or indeed the layman.

Introduction

1

In this introductory chapter, floods are defined and the overall national political and insurance situations are considered.

Floods have always been a feature of the landscape in Britain, but the floods of the past few years have considerably raised the national awareness of flooding and flood risk. Discussion of floods and their effect on property frequently receive extensive media coverage. With the aid of this book, the reader should be able to form a fairly accurate assessment of flood risk for any single building or site.

Definition of floods

Most readers will probably have a reasonable idea of what constitutes a flood. A quick investigation reveals the following: "an overflowing or influx of water beyond its normal confines, especially overland; an inundation" (*The Concise Oxford English Dictionary*) and "a high level of water which overtops either a natural or an artificial riverbank, or rises above normal levels or sea defences in coastal environments" (*The Oxford Companion to the Earth*). Floods, then, may be of marine or riverine origin.

The significance of floods

According to insurers Norwich Union (Aviva), approximately 10% of the population in England live within an area that is potentially at risk of flooding (ultimate source: DEFRA).

Between 950,000 and 1.2 million properties are built on inland floodplains in the UK (source: ABI). The Environment Agency website states: "Around 5 million people, in 2 million properties, live in flood risk areas in England and Wales".

The Office of Science and Technology report, *Foresight Future Flooding*, states that: "nearly 2 million properties in floodplains along rivers, estuaries and coasts in the UK are potentially at risk of river or coastal flooding. 80,000 properties are at risk in towns and cities from flooding caused by heavy downpours that overwhelm urban drains — so-called 'intra-urban' flooding. In England and Wales alone, over 4 million people and properties valued at over £200 billion are at risk".

The same report quotes the average annual damage cost in England and Wales from river and coastal flooding at £1,040,000,000, while the costs of flood management in 2000–04 are quoted as £439m. Around 1,740,000 properties are said to be at risk in England and Wales (figure 1). Figures quoted for Scotland and Northern Ireland are less dramatic. These figures almost certainly exclude all sorts of incidental costs due to problems with diversion of transport and such like.

Figure 1: Open all day — to canoeists (photo by R T Edwards)

The same report says, "the estimation of future flooding risks is difficult due to future uncertainties. However, all scenarios point to substantial increases".

A press release by the Benfield Hazard Research Centre (University of London) in July 2004 states: "Professor Bill McGuire, Benfield Professor of Geohazards and Director of the Benfield Hazard Research Centre says: 'There is now no question that hothouse Earth will be a far more hazardous place, with extreme meteorological events — in particular *floods* [my italics], drought, wildfires and heat waves — becoming more and more commonplace. In the year 2000, 1 in 30 of the global population was affected by natural hazards. By 2100, it is not impossible that climate-related natural hazards will impinge upon the lives of everyone on the planet'".

A report by the Royal Institution of Chartered Surveyors in September 2004, based on a survey of valuers across the country, suggested that properties in flood risk areas would typically be downvalued by 12–15%, and in some cases by 40%. The report emphasises that the effect on individual properties may vary greatly.

Since 2000, colossal sums have been expended by insurers on British flood damage. This has led to a situation in which properties in flood risk areas may no longer be able to obtain cover, or may suffer vastly increased premiums.

Insurers have reacted to both recent flood disasters and to forecasts of worsening future problems in different ways. Norwich Union has generated its own digital elevation model for England, Wales and southern Scotland, based on airborne radar mapping. The company has combined this with a software model that produces information on where floods will likely occur, how they could spread and how far they could extend. Norwich Union hopes that this will increase the number of properties that can be excluded from flood risk for insurance purposes in the areas that were previously assessed, crudely, on a postcode basis. However, they have also warned that properties that are built on insufficiently protected floodplains may be unable to achieve cover.

Many of the areas within the proposed Thames Gateway developments are seriously at risk, and insurers have issued warnings to the government.

Meanwhile, the Environment Agency has upgraded its website dedicated to flood risk prediction, which has been running for several years. The new system shows areas of potential river and marine flooding, distinguishing separately those areas that are thought to be at risk only during times of exceptional flood. The latter are defined as having a chance of one in 1000 or less of being flooded in any given year. Reports indicate that as soon as the revised site went online, the servers were besieged.

The site is full of useful information. The Environment Agency states: "Our aim is to reduce the risk of flooding to people's lives, homes and businesses and to the country's infrastructure and environment. By 2007 we will have reduced the risk to 80,000 homes.

We can do this by:

a. ensuring more residents in flood risk areas know they are at risk and take action to protect themselves
b. improving the coverage of our flood-warning services
c. reduce the number of properties exposed to a high risk of flooding
d. plan our defences based on how a whole area is affected, rather than individual locations
e. prevent all inappropriate development inside floodplains

Once it is realised that we can't stop floods from happening, the next logical step is to tell people how they can protect themselves. There are steps everyone can take to reduce the impact and danger of flooding, which you can find on our website or obtain through our 24-hour Floodline service."

The website provides ancillary information (for instance, the rights and obligations of riparian owners, and a flood guide for businesses), which can be discovered by a diligent search of the many flood-related pages. The flood risk maps are covered in chapter 6.

The above information, which has probably been superseded by the time that you read this due to the rapid rate of re-evaluation, should be enough to convince even the most sceptical surveyor or other building professional that knowledge of flood risk, flood management and risk assessment techniques are of great importance. This book sets out to give you the necessary practical tools.

In part, the greater risk that we now perceive is false. The risk has always been present. The landscape is not as stable as most people believe. On the scale of a human lifetime, matters may seem stable, but research demonstrates that climate is extremely variable. Between 1500 and 1850, a period which is known to scientists as the "little ice age", temperatures in this country were on average significantly lower, so that it became possible to hold "Frost Fairs" on the solidly frozen river Thames in central London. Additionally, global sea level rises and falls according to the amount of ice locked up in polar ice caps, while the earth's crust itself tips gently this way and that, gently rising and falling in different areas, under the influence of immense geological forces. It is the slow scale of change that deludes us into believing that nothing untoward will happen.

At the time of writing, all the indications are indeed that global climate changes will lead to rising sea levels and, in the British area, increasing storminess.

River and Stream Flooding

2

An introduction to the behaviour of rainwater, groundwater and rivers.

Basic hydrology

Hydrology is the discipline that studies water, its properties, and the way in which it is circulated through the earth and its atmosphere. As with many scientific disciplines, it is a synthesis of many other disciplines.

The "hydrological cycle" or "water cycle" is the term used to describe the way in which water endlessly circulates from the atmosphere and oceans to the land and then back to the oceans. In brief, the sun's radiation causes water to evaporate from the sea and from lakes and rivers into the atmosphere, where the water vapour may condense as clouds, which then form rain, depending on meteorological conditions. It is estimated that 97.2% of the earth's free water is in the oceans. The remaining 2.8% includes both fresh and slightly salty (brackish) water. Of this small proportion, approximately 77% is trapped within ice sheets and glaciers. Some 22% lies just below the surface of the earth as groundwater, while the remaining 1% is in transit through the atmosphere, rivers, lakes and superficial soil.

In short, the damage that floods cause is due to a small percentage of the earth's water. The way in which this tiny percentage behaves has been minutely studied but, nevertheless, models of rainwater and river behaviour are still far from perfect.

What happens to rainwater

Rainwater striking the earth's surface may immediately run off into a watercourse and head for the sea or a lake. Alternatively, it may soak directly into the soil. The properties of the soil, the rocks underneath, and vegetation in the area then govern its behaviour. Water that is not speedily taken up by vegetation may pass through the soil into a nearby watercourse, or it may penetrate into the rocks beneath. These rocks may be variously permeable (able to transmit water) and / or porous (having pores capable of holding water).

Rocks that are capable of transmitting significant quantities of water are termed "aquifers". Rocks that do not transmit significant quantities of water are termed "aquicludes".

Rainwater that travels over the earth's surface becomes organised into watercourses. The watercourses form a network that evolves continuously, but slowly, and is normally "dendritic" (a map of the watercourses resembles a drawing of a tree). Smaller channels merge to form larger channels, which, in turn, coalesce to form still larger channels. These channels, as they become established, cut into

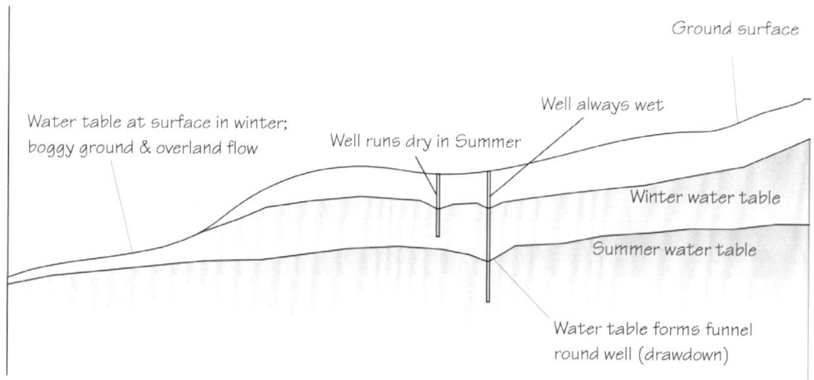

Figure 2: Schematic cross-section showing seasonal changes in water table

their substrates by erosion and deepen. Once they are established, in the absence of other changes, any given drop of rainwater, which falls at a particular point and does not soak into the ground, will always run over the surface towards the same channel.

When rain is not falling, rivers still flow. This is because groundwater continues to feed them. Groundwater may feed directly into a main channel, or into the finest of tributaries. In order for these channels to carry water, the water table must be at the ground surface.

The water table (figure 2) is a notional surface. It may be best imagined by supposing a near-infinite distribution of wells. The water table at any given time is the imaginary surface that unites the water levels in all of these wells.

A water table typically shows a form similar to that in the diagrams but, in reality, it may be much more complex as it is governed by a variety of features, such as differences in permeability of the rocks, the presence of aquicludes, borehole extraction points, cave systems, railway tunnels, dams and so forth. It is important to understand that aquifers may be extremely dynamic and that changes in rainfall, water abstraction from boreholes, or other influences may have rapid effects.

Rainfall "recharges" an aquifer in its "recharge area"— that is, in general, the area where the aquifer outcrops at the surface. For instance, the Chalk aquifer under London is recharged in the North Downs and Chilterns. To some extent, leakage from watercourses and lakes may also recharge aquifers if the water table is low at such locations.

Where the water table meets the ground surface, water will begin to flow. After periods of persistent rain, the water table generally rises overall. Whether this water immediately forms a channel depends upon a number of factors, including whether there are any pre-existing channels in the vicinity. Normally, there will be, because of the maturity of the British landscape. Under some circumstances, water may merely ooze out over the surface to form a "flush". Such locations are normally boggy on a semi-permanent basis. Where these areas are not being managed, they normally support the growth of rushes.

If rainfall is particularly heavy and / or persistent, the water table will rise still further, and yet more water will come out at the surface and flow overland to the nearest watercourse.

If the water emerging from the ground is concentrated at a particular point, this is normally known as a "spring". Springs may be more or less permanent, and dry up only in the most persistent drought, or they may be intermittent if they are not regularly reached by the water table.

The location of any particular spring will be influenced by a number of factors but it is common for them to be aligned in an obvious pattern. This is normally because of a change in permeability of the underlying rock. In lowland Britain, the lines

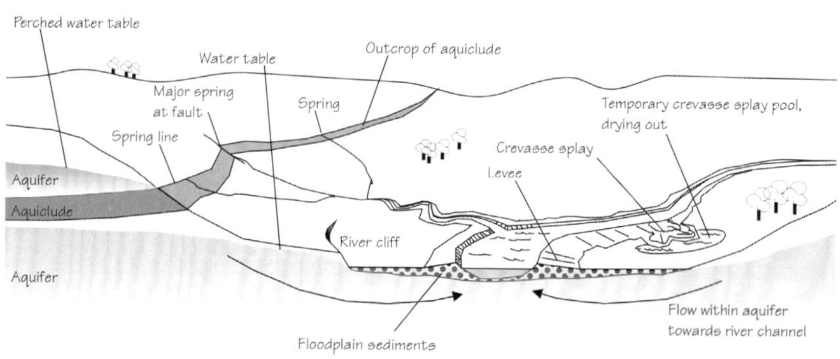

Figure 3: Cross-section through floodplain showing aquifers

Figure 4: The Lewes Winterbourne flows after a wet autumn

Figure 5: The Wellesbourne flows down Old London Road, Patcham

of springs commonly follow the outcrops of clay strata, which are generally more or less impermeable. Where, as is usually the case, there are ranges of hills, such as the Downs or the Chilterns, which comprise permeable strata (in this case chalk) overlying impermeable strata, it is common to find a row of villages or hamlets, each focused on a spring: the classic "spring-line settlement" beloved of geography teachers (figure 3).

Where rainfall is sparse, on the contrary, rivers may flow intermittently or seasonally, or (in the extreme case of a desert) not flow for many years on end. This is because the water table is normally below the level of the river bed (or at least the bottom of the valley). In Britain, this phenomenon occurs commonly on the limestone and chalk uplands, where the water table is normally below the surface except after persistent wet weather. Matters have been complicated by the extraction of water from boreholes, which lowers the water table still further. However, the impersistent nature of such streams has been recognised since time immemorial, and on chalk topography in South-East England such streams are commonly termed "bournes". The element "bourne" commonly occurs as part of a longer name, for instance Winterbourne and Broxbourne. "Winterbourne" (figure 4) is almost self-explanatory — it signifies a stream that flows in winter only. Some winterbournes flow so rarely that they may be built over, with unfortunate results. The Wellesbourne, in Brighton, had not flowed for over a century (due largely to water abstraction from the chalk). However, the exceptionally wet year of 2000 caused it to reappear, flooding cellars, basements, shops and homes in the suburb of Patcham (figure 5).

Even quite large streams or rivers in South-East England can disappear during dry periods, so that the River Darent at Dartford may be extremely low, and indeed on several occasions has dried up in its middle course. The River Mole, north of Dorking, often disappears, in this case into a subterranean drainage system within the chalk. In areas such as the Peak District, parts of the Pennines and parts of South Wales, there are Carboniferous limestone outcrops. These areas are virtually devoid of surface streams, since the limestone is conducive to the development of extensive cave systems into which rainfall disappears almost immediately.

The behaviour of the water table is not as simple as it might appear from the previous paragraphs. The distribution and form of aquifers may be quite complex. The overall pattern of the geology of most of southern and eastern England, together with parts of the Scottish Lowlands, comprises alternating layers of permeable and impermeable strata. However, the "layer-cake" arrangement is complicated in detail. Aquifers may be cut off laterally by folds and faults. They may "pinch out" as a result of original non-deposition of strata in a given area. They may terminate against igneous intrusions, or may be divided into compartments by narrow igneous intrusions, such as dykes or sills, or by faults. They may even be "perched" (figure 3), where high ground, which is mostly permeable, features a relatively thin, sub-horizontal impermeable bed.

Dams and railway embankments, motorway cuttings and other engineering works all affect the behaviour of aquifers. If you have been into a cave or mine tunnel in the UK you may remember how water usually drips from overhead. This is because you are, essentially, inside a drainpipe cutting through an aquifer. The removal of groundwater from mines is a major consideration in most mining operations.

The performance of an aquifer is affected by its porosity (the percentage of void space) and its jointing and / or fracturing network. The chalk, which is typically highly porous (about 35%) has such small pores that the rate of water transmission is low, and most of the water flows through a network of fine fractures. The hard limestones of the Carboniferous have very low porosity, but have developed a system of broad joints and fractures, which have been enlarged by dissolution (including the formation of major cave systems).

Without straying too far into the realm of engineering, it is worth noting that water table levels can affect the engineering properties of the rock beneath a property. During the period of large-scale abstraction from the aquifers beneath London in late Victorian times and much of the 20th century, the consolidation of dried-out rock strata led to about 300mm of subsidence in central London. The bearing capacity of piles is reduced in saturated subsoils, while pressures on basement walls increase if the water table rises. The reader should be well aware of the effects on shrinkable clays of changes in moisture content.

In upland Britain, aquifers are usually poorly developed, since apart from the Carboniferous limestone areas, the bedrock is usually hard and dense and has poor permeability. Here, the importance of superficial deposits is much greater. Such deposits are generally either riverine alluvium, or sands and silts deposited as a result of glacial action. These deposits are usually thin and have only limited capacity to hold water.

Upland Britain is also noted for the localised development of peat, which has a substantial capacity to hold water. Where peat deposits remain, rainfall is prevented from running off immediately and is released slowly by the peat. However, once the peat has become saturated this buffering action is lost, and the rainwater runs straight off into watercourses.

Remember that a watercourse is only the surface manifestation of the movement of water. Flow also occurs — although much more slowly — through the sediments and bedrock beneath the channel. In fact, it occurs everywhere below the water table, although some aquifers are almost static, with water residence times of tens of thousands of years, or more. River and stream channels do not have "bottoms" in the same sense that a rainwater gutter has a bottom. They are in hydraulic continuity with the aquifer beneath them.

Even at this basic level of explanation, one can see how difficult it is to predict the behaviour of groundwater in any given area. Predictability may be enhanced in areas where there is a great density of wells or boreholes with monitoring devices, but even the best computerised hydrological models are often defeated.

The development of floodplains

Although geomorphologists are no longer quite as complacent as they were in the 20th century, there is still consensus that in areas with reasonable rainfall (including Great Britain) the development of watercourses follows a recognisable pattern. This pattern can be modelled, crudely but effectively, by piling up fine sand. If an overall fine but heavy spray of water is then applied, from a hosepipe with a sprinkler, a stream system will soon develop. Gullies develop in the upper part of the sand pile as the moving water picks up the loose sand, and the eroded sand is then redeposited at the margin of the heap. This takes place over a period of only minutes, and in a medium that is virtually uniform. In reality, matters are naturally more complex.

Suppose that a landmass is emerging from the sea (for instance, the volcanic island of Surtsey off Iceland). As soon as rain falls, some of it runs off and well-defined channels rapidly develop. These channels are not straight, although the steeper ones may be nearly straight. There is a natural tendency for the water to oscillate from side to side as it flows, and this rapidly produces curves, or "meanders" in the stream channel. These meanders are not static; they change their position in a downstream direction. This is the basis for the development of a floodplain.

Where the land is rapidly rising, uplift means that the stream channels must be very steep. In consequence, their work is almost completely erosive and the channels cut rapidly into the emerging landmass. However, where matters are proceeding in a more leisurely fashion, sediment may be picked up by the stream and redeposited fairly rapidly, depending upon fluctuations in rainfall. The redeposition can

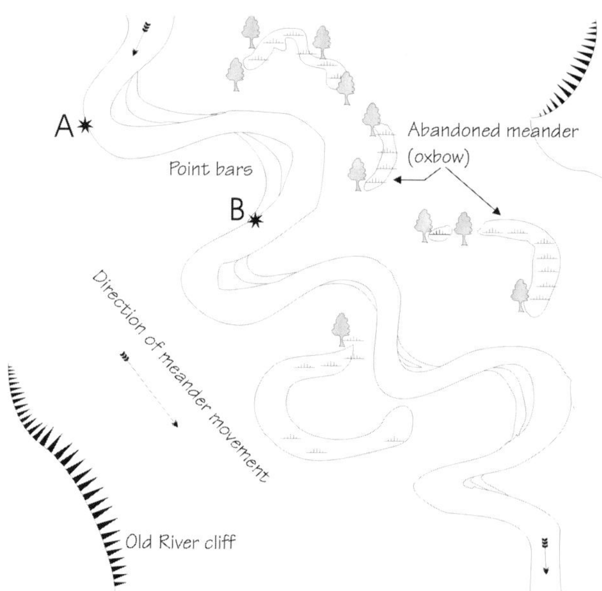

Figure 6: Movement of meanders across a floodplain

Figure 7: The River Arun occupies its floodplain at Pulborough

Figure 9: High water table shows up sections of former watercourse

Figure 8: The River Severn in flood near Upton (photo by R T Edwards)

take place only within the valley bottom, which the stream has already cut. This area, in which erosion and redeposition takes place, is the floodplain.

In a relatively mature landscape, such as that of the British Isles (although the British Isles are virtually newly exposed after the last glaciation, compared with an area such as the Amazon basin), all these processes have been operating for some time. The upland areas in the northern and western parts of the islands are generally characterised by steep valleys with little or no sediment deposited in the bottom. In the lowland areas, one finds broader valleys, often with substantial spreads of sediment, which are crossed by well-established trunk streams and rivers with well-developed meanders.

The meanders gradually migrate downstream (figure 6). This is due to the erosive power of the river, which erodes the downstream sides of the meanders and almost immediately redeposits some of the sediment on the upstream side of the next meander. In figure 6, for example, sediment eroded from point A may be redeposited temporarily at point B. Consequently, the sediment also gradually migrates towards the sea. Any given particle of sediment, however, may take thousands or tens of thousands of years to complete its journey. A section cut into the floodplain (figure 3) will show several metres, or tens of metres, of silts, muds, sands and some gravels, together perhaps with thin layers of organic material, such as peat. Due to land reclamation, peat is now not a normal constituent of lowland river valleys in the well-populated parts of the British Isles. However, valleys in western and northern Britain still contain significant lowland peat deposits.

In times of exceptional rainfall, the river ceases to flow only within the channel and occupies the entire width of the floodplain (figures 7 and 8). It is at this time that the maximum erosion takes place. In areas that have not been affected by man-made river defences, bridges and other artificial features, it is quite possible for the river to change its course completely and dramatically, almost overnight. As the water level falls again, a new topography may be revealed. During the process of the flood, substantial amounts of sediment will have been moved. Figures 9 and 10 give an idea of how the volume of water increases during a flood and makes such work possible. The distinctive tree group shows that the same bridge is involved.

The floodplain deposits are essentially impermanent. A floodplain is an area of temporary storage for river or stream sediment. For man to develop settlements on the floodplain, and then to complain if they are washed away, is hubris. This is, however, understandable when

Figure 10: Bridge — winter low water

Figure 11: Bridge — winter flood

one compares the puny lifespan of a human being with the frequency of major floods. The floodplain surface may not seem temporary in the context of a human lifespan, but we have to bear in mind that it can be changed overnight. This lack of understanding is the root of most floodplain problems.

As rivers or streams fill to the brim, this is termed the "bankfull" stage. At such a point, the channel is usually full of sediment-laden water. As the water level increases to the point where the channel overflows (the "overbank" stage), water that leaves the channel experiences a loss of velocity, which causes it to deposit some of the sediment. This sediment then builds up at the edge of the channel like a natural embankment (figure 3), which is termed a "levee" (from the French *levée*, meaning "raised"). This term is not normally used in Britain, but is associated particularly with the river Mississippi in the US, where the term was applied in the French-speaking areas around the lower reaches of the river.

As meanders loop irregularly, sections may be cut off and turned into backwaters, then into shallow, elongated pools (oxbows) that gradually silt up. However, they may remain for a long time as hollows that fill with water during times of high water table (figure 9).

Natural levees do occur in Great Britain, but usually only on relatively minor stream channels, since elsewhere they have long ago disappeared under artificial embankments and other man-made works.

Natural levees, once formed, act as a natural defence to the areas of floodplain behind them once they have become stabilised. However, they are vulnerable to breaching, and when they are breached the river deposits a splay of mud, silt and sand over the area of floodplain immediately adjacent to the breach. This is one of the processes by which the floodplain is built up before the next sweep of a major meander erodes it away and redeposits the sediment further downstream.

Figures 12 and 13 show artificial levees.

Once the levee has been breached, mud-laden water tends to become trapped behind the remaining levee (figure 13) and can take a considerable period of time to dissipate. In urban areas, this may necessitate pumping by the Fire Brigade. The mud, which was held in suspension in the moving water, is deposited and may

Figure 12: Even at normal high tide, the river level is above the surface of the fields

Figure 13: The River Ouse in flood below Lewes

amount to centimetres or more in thickness. This can contribute substantially to property and other damage.

The repeated redeposition of sediment over floodplain areas is one of the ways in which natural soil fertility is maintained, and why the annual flood of the River Nile was, in the past, so crucial to the prosperity of Egypt. Now that flood management systems (as discussed below) depend increasingly on allowing natural flooding to take place, dairy herds can take advantage of the lush growth that follows the flooding of water meadows.

Where the river nears the sea, it may also be liable to marine flooding, or a high tide may hold back river water, which consequently is forced to overflow.

River catchment areas (the area of land that drains, ultimately, through a particular river mouth) are far from uniform (compare the maps in the case studies in chapter 5). In practice, some river basins develop at the expense of others, because of various factors, such as differing bedrock resistance. These larger basins inevitably result in the lower reaches of the river becoming a bottleneck, so that unusually high rainfall over the catchment results in a disproportionate rise in water level.

Different flood types

Flash floods

The term "flash flood" has been rather overused in the press, and has been applied, often indiscriminately, to any localised and catastrophic flood. Unlike the relatively predictable floodplain flood, flash floods result from a sudden influx of water over a very short timescale, often a matter of hours. In the case of the Boscastle flood in 2004 (figure 23 on page 22), there was virtually no warning, although persons in the affected area had time to become concerned. Because of the rapidity of the onset of such floods, loss of life is a much greater risk than with floodplain flooding.

The essence of the flash flood is that a concentrated amount of rainfall falls in a relatively short timescale over an area where the physical characteristics promote such a flood. Predisposing factors include steep topography, absent or sparse vegetation, hard surfaces such as bare rock, car parks etc, and features such as narrow valleys, bridges or culverts, which can channel or restrict the resultant flow.

In reality, there is no hard distinction between the various types of flood: for instance, a flood caused by the breaking of a levee might be so sudden as to be termed a flash flood. The failure of artificial features, such as dams, can also give rise to such flooding.

The ability to forecast flash floods is limited because the relationship between the rainfall and the topography is not easily recognised, although long-term residents in a given area will probably recognise the dangerous type of weather.

Sheet floods (or surface water run-off)

In some cases, rainfall can be so intense or prolonged that it flows directly overland as a significant depth of water. It can occur over limited areas between stream channels and I am aware of properties that have suffered in this way. Fortunately, this does not occur on a large scale in the British Isles.

Groundwater flooding

Where the water table rises so that it is notionally above ground level, a flood will by definition occur. However, this may be completely masked by other forms of flooding. In practice, the most obvious case in which the water table rises above the surface is where seasonal streams (as described above) begin to flow. However, these are then usually confined to valley bottoms.

Global warming

There cannot be many readers who are unaware, at least broadly, of the effects of global warming. There is sufficient evidence to suggest that there is a real and significant effect, which shows itself readily in the thinning of the Arctic icepack and a statistical rise in temperatures. Many earth scientists suggest that this will give rise to persistent and increased storminess, which is likely to result, in turn, in a greater incidence of intense rainfall. Whether this global warming is the result of man-made factors, or is merely a natural variation, is unknown. Scientists who study esoteric matters, such as oxygen isotope ratios, fossil tree pollen etc, have known for a long time that temperatures have fluctuated, often quite wildly, even over the last 10,000 years. Whichever scientific explanation may be correct, the point is that we cannot expect stability.

Controlling and forecasting floods

Flooding is probably the most widespread of all geographical hazards. Floods are responsible for major economic loss. While the level of mortality from floods is below that of other natural hazards, such as earthquakes, drought and famine, the level of homelessness following floods is very high. Because of this economic loss, authorities have for many years attempted to limit the damage by various forms of control. To begin with, starting in the 1930s, engineering solutions were sought, including the construction of large holding reservoirs, the construction of levees and embankments, and through channel "improvements" (usually, widening and / or straightening). However, major watercourses are not particularly amenable to all engineering solutions. A river that is forced to flow in a straight line will not always co-operate. It attempts to meander within its straight channel, and deposits alternating bars of sediment to either flank. The discharge that is removed from the relevant area may arrive prematurely downstream and cause problems elsewhere. The deposition of sediment on the bed of the channel may simply result in overtopping of the expensively constructed channel.

Sediment that is normally carried away by the natural flow regime of a channel will be trapped behind a dam and soon fill up the reservoir. The loss of sediment to the downstream part of the channel may itself result in changes in flow behaviour.

In the 1960s, a different approach was tried. Although some reliance was still placed on heavy engineering solutions, a "soft" approach included a combination of other techniques. Greater reliance was placed on instrumentation, such as flow meters, telemetry of data, computer analysis and meteorological forecasting. The planning system was increasingly used (although not consistently) to control property development in areas prone to flooding, while insurance was used as a technique to offset costs.

There is now a considerable legacy of older river defences in place, of varying effectiveness.

The present approach (although, again, there is a lack of complete consistency) is to attempt to restore rivers and floodplains back to their natural state. Areas of floodplain may be reserved as "washland", which are allowed to flood freely. This absorbs water that would otherwise be surging downstream, and delays and smoothes the peak water level. Afforestation of bare slopes reduces initial run-off and landslide risk.

In areas that have already been developed, this approach is not usually a viable strategy.

Defence schemes can be expensive: West Dorset protected just 57 homes in Piddletrenthide at a cost of £1.7m by boring a bypass culvert (completed in 2005) to divert water from the River Piddle through a hillside and back into the river downstream.

Quite recently, the development of improved flood forecasting techniques has been linked with the ability to transmit warnings more readily. On the defence side, considerable success has been experienced with demountable barriers at such locations as Bewdley and Ironbridge in the flood-prone Severn valley. Forecasting and response times are naturally critical for the early deployment of such defences.

Increasing areas of Great Britain have disappeared under hard surfaces, such as high-density housing estates and retail parks with large car parks. New home owners often lose no time in adding further to the amount of hard surface, while industrial occupiers may also extend parking areas. The buildings themselves are, of course, also hard areas. All these hard surfaces prevent rainwater from soaking into the ground. Rainwater that would naturally soak into the ground and recharge any underlying aquifer is forced to run over the hard surfaces, into the storm water drainage system, and thence rapidly into the natural drainage systems. Few developments, as yet, include permeable surfaces.

Where defences exist, if they are overtopped or breached they may impede the subsequent drainage of the floodwaters, and pumping may be necessary.

Perception of risk

In the mind of the British public, the perception of flooding has been greatly altered by the headline-grabbing major floods of recent years, accompanied by dramatic television footage. There has also been a trend towards better flood-warning systems, for instance the prevalent warnings given after adverse weather forecasts. Another factor that highlights perceived flood risk has been the availability of the latest version of the Environment Agency online flood maps.

It is doubtful, however, whether this level of perception reflects a real change in flood risk. Although, as I have said, global warming does appear to be a reality, the timescale that we are forced to consider is too short to be certain that we are not dealing with events that are partly random.

For the surveyor, the risks are seen in terms of valuation, desirability of the property, and Professional Indemnity risk.

Marine Flooding

3

After river flooding, flooding by the sea is the major risk. The common factors are examined in this chapter.

Any land that is below the potential maximum sea level at its location is axiomatically at risk of flooding.

As we all know, the sea level varies daily, with high tides approximately twice a day. The tides are caused by the gravitational effect of the sun and the moon. Tide attitudes vary greatly according to the size and shape of an ocean basin and its marginal inlets. In particular, estuaries have a large tidal range. In Britain, the estuary of the River Severn is generally taken to have the highest tidal range (ie between high and low tide levels) of about 12.3m.

The height of a high tide varies substantially during the lunar month. Neap tides are the lowest, and occur when the moon and sun are at 90° apart in the sky. At such times, the gravitational effects of the two bodies are partly cancelled out. Spring tides are the highest, and occur when the moon and sun are either aligned as seen from the earth, or are in opposition (ie the moon and sun are on diametrically opposite sides of the earth). At such times, the gravitational forces are at their maximum.

The very highest water level, of course, is not that of the notional highest tide, even when driven by a storm surge. It is the level of any waves that are superimposed upon the highest tide. Figure 14 (while not of a coastline at risk of flooding) gives some idea of the tremendous amount of energy expended against a coastline during a storm. Once the waves break through, it is difficult to prevent disaster.

Fortunately, in this country, there are only a few areas that are actually below mean sea level; most of these are in the Fens. However, there are substantial areas that are below high tide level, and the risk to these areas will vary according to the state of the tide at any given time. Such areas, once flooded, may take a very long time to dry out, even if assisted by pumping.

The risk from marine flooding is influenced predominantly by two factors: the condition of the defences, which is directly under human control,

Figure 14: Waves pounding the cliffs at Saltdean

and the condition of the sea, which is far from under control. Obviously, a set of sea defences that are poorly maintained may be overwhelmed by a storm of relatively moderate intensity, while a brand new, well-designed set of defences may be proof against all but the most catastrophic storm surge.

In practice, many lesser factors combine to affect the two main factors.

Sea defences

Natural defences

Substantial areas of the British coastline are protected to a certain extent by natural defences — "barrier beaches" (figure 15). These natural defences are, of course, not usually continuous, as they must be breached where rivers and streams flow to the sea. However, where the stream is not powerful and cannot maintain its channel, the barrier beach may be constructed across its mouth, with the result that the stream may escape to the sea through seepage. Such beaches are developed under circumstances where storm waves drive beach material up slope to a point above normal high tide level.

These circumstances develop under a complex combination of factors. In simple terms, in stormy weather, the waves thrust sediment up the beach. As each wave retreats its energy is dissipated by drainage into the sediment and other factors, and as it withdraws it carries the finer sand, silt and mud back. However, it is not energetic enough to transport the coarser sand, pebbles or cobbles, which remain at the head of the beach. To allow the formation of a barrier beach, there must be a supply of sediment, and suitable dynamic conditions. A detailed study of these is beyond the scope of this book.

The sediment from which the natural defence is composed may vary in size from large cobbles to the finest sand or silt. Probably the finest example of the former in the British Isles is Chesil Beach, which stretches from West Bay in Dorset to the Isle of Portland. Pebbly or sandy barrier beaches are more common, and may grade into dunes. Dunes are formed where sand, particularly fine sand, is redistributed by the wind. Dunes can reach great heights: the highest dune system in the British Isles is thought to be that at Merthyr Mawr Warren, in south Wales, which tops 180m.

Barrier beach and dune defences can provide local protection against overtopping by storm waves or surges. They remain more or less stable because they absorb the energy of the sea, although changes in the marine dynamic environment can result in the rapid disappearance of the barrier. The classic case

Figure 15: Here the natural barrier beach has been artificially regraded

of this is the destruction of the village of Hallsands, in Devon. This village stood at the foot of low cliffs, and was protected by a pebble beach from the sea. Despite the fears of the local inhabitants, a contractor was permitted to dredge offshore for gravel for construction. Within the space of a few years at the end of the 19th century, the village had been badly damaged and by 1917 it had all but disappeared, despite the construction of a sea wall. The dredging had resulted in a critical alteration in the circumstances that permitted the pebble beach to remain.

Where such natural barriers are not continuous, they will not provide protection against sustained onslaught, such as by a storm surge. The sea will, sooner or later, find its way around the end of the barrier.

Natural defences can, however, be augmented by man-made works. Small streams may be bridged over and closed by sluices. Where larger channels penetrate the sea defences, artificial embankments may be continued inshore along the banks of the river. In such cases, the risk of flooding is apportioned between marine and fluvial causes.

Artificial sea defences

Large areas of the UK have been reclaimed by constructing dykes or embankments at the seaward edge of land, which would normally, and naturally, be flooded at each high tide. In many cases, such embankment construction was repeated several times, and it is possible to trace, on the map, the successive lines of defences.

To begin with, the low-lying land behind the defences was drained by windmills. Later, steam engines were used, but these have long since given way to electric pumps. This arrangement depends upon constant vigilance and repair, since the breakdown of one element of the system could result in catastrophic failure and flooding. In areas such as the Fens, management of the (now largely artificial) river channels is an extremely complex matter, since the flow from different river channels will vary according to the nature of their catchments, local weather patterns, and their individual characteristics. The flow must be balanced so that it does not arrive as a peak.

Shrinkage

Where sediment has been dried out by artificial drainage, it shrinks. This is especially the case where peat deposits are present. In the Fenland, Whittlesea Mere was drained in the mid-19th century. A post (Holme Fen Post), which was erected at the time, recorded a surface level of 1.74m above the ordnance Datum at Newlyn in 1848. Since then, the surrounding peat has shrunk, and the surface has dropped by about 4m — ie significantly below mean sea level. Such areas, if their defences are breached, are liable to deep inundation.

Storm surges

Normally, the greatest risk of marine flooding will be at high tide. The level of the high tide varies, not only according to the positions of the moon and sun, but also according to the weather. The direction of the wind and atmospheric pressure can raise or depress the level of the water. The effect is more marked in shallow seas, such as the North Sea. In the case of the infamous 1953 storm surge, gusts of wind exceeded 100 knots and atmospheric pressure dropped as low as 968 millibars. This combination added 2.2–3.0m to the predicted tide level.

At the time of the 1953 surge, the recurrence interval of such an event was calculated (statistics again!) as 200 years. However, the 1953 level has since been exceeded in eastern England a dozen times. Research on Dutch sediments suggests that there has been a progressive increase in the height of extreme levels (presumed to be associated with storm surges) in the last 2500 years. The present Thames Barrier is a sign of the recognition of the risk.

Sea level changes

Sea level is rising globally. The Intergovernmental Panel on Climate Change undertook a review of published records in 1995. It reported that global sea levels had risen by 100–200mm in the previous century (ie, about 1–2mm per year on average). Slowly but inexorably, the strain on sea defences is increasing. Superimposed on this, there are local changes in sea level. In north-west Britain, particularly western Scotland, the earth's crust was depressed by the weight of ice during the Ice Age. When the ice melted, the crust began to recover (a process known as "isostatic rebound" — figure 16) and this continues slowly, so that old beach lines can now be recognised well above sea level in many areas. The rise in western Scotland is estimated to be about 1.6mm per annum.

In the populous areas of eastern and south-eastern England, by contrast, the land is sinking relative to sea level (figure 16). It is unclear to what extent this is due to a seesaw-like effect as the north-west rises, or to continuing subsidence of the North Sea basin, which has been more or less continuously subsiding and accepting sediment for hundreds of millions of years. Roman occupation levels in parts of London were at least 2m below modern high tide level, and the Thames was apparently not tidal at London. Today, tides extend to Teddington Lock, on the western fringe of Greater London. It is thought that tidal range has been partly increased by the extensive dredging prior to construction of the London docks during late-Victorian times.

Tide gauges and other data in the lower Thames area indicate actual rises in mean sea level, which

Figure 16: Crustal movement of the British Isles

are variable, but are perhaps typified by 2.14mm per annum (with an error range of plus or minus 0.15mm) for Sheerness, Kent, over the period 1901–96. Other figures from Greater London storm surge and tide gauge records suggest a rise of 8mm per annum. However the figures are interpreted, what is certain is that sea levels are rising relative to the land. As far back as 1928, a storm surge flooded parts of central London and drowned 14 people, and the city had a narrow escape in 1953; the risk is now greater. Large areas of London rely upon river defences.

At the time of writing, the Environment Agency is reported to be seriously examining a report, prepared by a group of academics, which envisages the possibility of a huge barrier that would cross the Thames Estuary from the Isle of Sheppey northwards to the Southend area, for possible construction towards 2100. The barrier would be similar to that constructed in the Netherlands to reclaim land. This proposal must be considered in the light of present intentions for massive development in the "Thames Gateway" area.

Perception of risk

As for river floods, media coverage strongly influences public perception. There is also an apparent lack of public understanding that low-lying coastlines can be massively altered in a short period.

Unusual Flood Hazards

4

It is easy to overlook causes of flooding other than straightforward river and marine high water levels. A few more obscure, but real, risks are considered in this chapter.

Dams

Much of Britain's water supply comes from reservoirs impounded by dams. Most of these are in rural and upland areas, since this is where valley shapes are suitable and rainfall is higher. In most cases, few properties are located in any downstream risk area; but the risk is there. Dams can be affected by design faults, terrorism and landslides. In fact, what is probably the world's worst dam disaster occurred in 1963, at the Vaiont Dam in the Italian Dolomites. Although the dam itself was not seriously damaged, a major landslide fell into the reservoir behind it, causing massive overtopping of the dam, and huge damage downstream. Some 2500 people were killed by the flooding.

As a related issue, the presence of a major dam can itself give rise to earth tremors because of the redistribution of stresses in the underlying rock. These tremors can cause landslides, which could cause flooding problems even if they do not involve the reservoir itself. Such landslides can form temporary dams and can impound enough water to cause serious problems if it escapes suddenly.

A 2002 DEFRA report showed that there have been no British dam-related deaths since 1925. Dam construction is now tightly controlled by legislation, so the risk is quite small. However, it is worth noting that of the 14 dam incidents that have caused fatalities in the UK since 1799, no fewer than nine involved overtopping rather than dam failure.

Mine bursts

Once abandoned, old mines normally become filled with water up to the level of the water table or to the level of any drainage adit (a roughly horizontal tunnel driven specifically for drainage to a nearby valley bottom). If such an adit, as often happens, becomes blocked by a roof fall, water can build up behind the blockage.

At Cwmrheidol mine, an abandoned lead-zinc mine near Aberystwyth, just such a blockage occurred and was disturbed by a JCB in the 1960s. Mine water laden with toxic metals poured out. In this case, about 2 million litres of water, laden with zinc, cadmium and aluminium was discharged (figure 17). Aberystwyth Harbour changed colour, and thousands of fish were killed — but, in this remote location, no buildings were damaged. In the 1990s, the same mine gave further trouble when mine subsidence diverted a stream into the workings, resulting in a further toxic efflux. As such, any development in locations downstream from such mining areas could be unwise.

Figure 17: Pollution from a mine burst (photo by J S Mason)

Tsunami

As I write, the death toll from the earthquake and tsunami of 26 December 2004 stands at nearly 300,000. The major earthquake and its aftershocks occurred under the sea off the western side of Sumatra, and while most of the tsunami damage occurred in Indonesia itself, western Thailand, southern India and Sri Lanka, damage was caused in Somalia and Kenya, which is over 5000km away. Few readers can have avoided seeing television footage of this dreadful disaster, and many will know of individuals caught up in it.

So what *is* a tsunami? The word is Japanese for "harbour wave". The term tidal wave is often erroneously applied: a tidal wave is what happens every day, and the two are not related. Earthquakes, underwater landslides, or both together can cause a significant displacement of water. Just as when a stone is thrown into a pond, the resultant wave pattern spreads out in a circle, and while it may be dissipated by obstructions, such as promontories, it is refracted around them.

In the recent past, scientific research has assessed two possible sources of such tsunamis, other than underwater earthquakes.

Gas hydrates are ice-like deposits of water and gas (mainly methane) that are locked up in ocean sediments, in cold high-pressure environments. As global warming proceeds, it is feared that the gas hydrate deposits will be destabilised and will break down, releasing their constituent components. This process has the potential to cause substantial underwater landslides, and these, in turn, could lead to tsunami formation on an unknown scale.

Cumbre Vieja is a volcano on the island of La Palma in the Canary Islands. It has been suggested that a section of the volcano, which is approximately the size of the Isle of Man, could suffer a catastrophic landslide following an eruption. This could lead to a devastating tsunami, which would have a substantial wave height on reaching the British coastline. Most low-lying areas would be at risk of overtopping of the sea defences.

While there has been much scientific criticism of these ideas, especially the Cumbre Vieja suggestion, the original scientific papers are based on extrapolation from existing knowledge and the risk should be considered real, if perhaps small. The terrible disaster caused by the South Asian earthquake underlines the vulnerability of man's activities to the unstoppable forces released by planetary processes.

Even in the British Isles, significant damage may have been caused by a tsunami in historic times. The serious flood of 1607 in the Bristol Channel and South Wales area took place in fine weather, and descriptions of the depth of water and rate of advance of the flood (faster than a greyhound) certainly suggest a tsunami. At least 2000 people were killed, houses were destroyed, and a laden cargo ship was beached well above high tide level at Appledore. The possible cause remains obscure, although an active undersea fault off southern Ireland may be to blame. The hypothesis was aired in a BBC *Timewatch* programme in April 2005.

Case Studies

5

Four well-documented modern British floods are examined and the main factors assessed.

Figure 18: Case study locations

Four significant floods have been selected. Their locations are shown in figure 18.

Lewes — 2000

The small town of Lewes is an ancient settlement, formed at a bridging point of the River Ouse in East Sussex. It is focused on the steep hill to the west of the river, but there is also significant urban development on the floodplain. At this point the floodplain is constricted by the chalk hill on which the town stands and by a further chalk hill mass to the east, which is separated from the main mass of the South Downs. At its narrowest point, the floodplain is less than 500m wide. The river channel is constricted by the development of the town and by two bridges close together: the old Cliffe Bridge and, to a lesser extent, the newer Phoenix Causeway.

The rural floodplain of the Ouse and its tributary streams are regularly inundated both above and below the town. Local land management takes account of this and few properties are at risk in these areas.

Lewes itself stands upon the chalk. However, much of the river catchment lies upon clay formations, while the higher ground to the north, in the centre of the Weald, is of alternating clays and sandstones, which are, overall, almost as impermeable as the larger areas of clay.

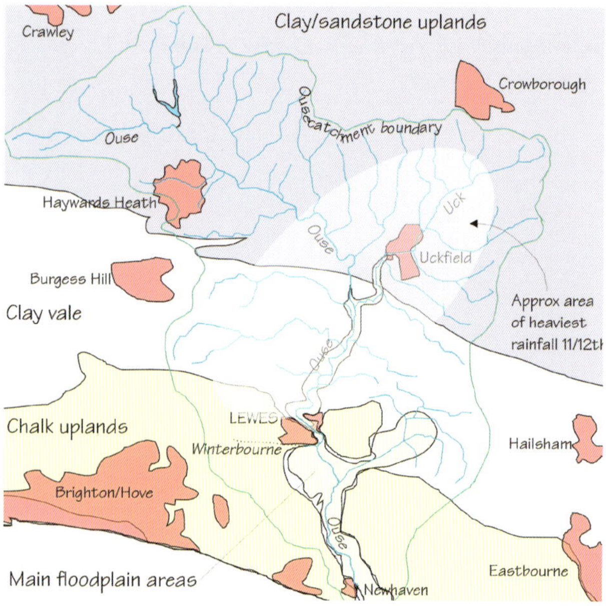

Figure 19: The Sussex Ouse catchment

Figure 21: The Lewes Winterbourne in its normal dry state

Figure 20: Harvey's Brewery was devastated by the Lewes floods (photo by The Argus)

October 2000 was very wet. The catchment of the Sussex Ouse was already saturated, and flood watch and flood warning notices had already been issued when particularly heavy rain fell on the night of 11/12 October. The heaviest rain was concentrated on the central Ouse catchment (figure 19) and the valley of the River Uck, a tributary of the Ouse. Because the ground was already saturated, river and stream water levels rose immediately, prompting further warnings. Early on 12 October, Uckfield, located on the Uck, flooded dramatically, with stock being washed out of shops (in some cases being found, later, miles downstream). As flood peaks from the Ouse and the Uck progressed downstream, the floodplain above Lewes filled up and water backed up behind the constriction formed in central Lewes. By early afternoon, flood defences were overwhelmed, the lowest part of the town centre filled with water, and hundreds of people were trapped. The Harveys brewery (figure 20) was under water, and casks floated on the flood. Flood levels peaked in the evening, and the low-lying Malling area was covered by up to 3.6m of water. Elsewhere, water depths of 1–1.2m were common.

The flood defences retained the flood waters as the river peak receded and flood victims were unable to return to their businesses and homes for two days.

During this period, the Winterbourne stream, which normally has a dry bed (figure 21), carried substantial quantities of water and flooding occurred when a culvert became blocked.

In Lewes, 836 properties were affected; of these, about 27% were residential. Damage to infrastructure was substantial.

In March 2001, the Environment Agency published a detailed flood report . This remarked particularly that the Ouse catchment is especially vulnerable to flooding, and that for centuries development avoided the floodplain areas. Development on the floodplain area in Lewes, including the construction of railway embankments (partly now disused) and industrial areas, is thought to have exacerbated matters by increasing surface run-off, reducing flood water storage capacity, and constricting the flow.

Below Lewes, flooding caused minimal damage.

The costs of the flood were estimated at £130,000,000, although some data was unavailable when the report was compiled, and the real cost will be higher.

York — 2000

York is another ancient settlement, but lies on generally low ground, and straddles the (Yorkshire) River Ouse. Much land adjacent to the river is 10m or less above Ordnance Datum. The river is formed by the confluence of the important Rivers Swale and Ure near Boroughbridge, and by the time it reaches York it has also received water from the major tributary of the Nidd (figure 22). All these rivers drain substantial upland areas of northern Yorkshire, with relatively high rainfall. At York, the Ouse is joined by the River Foss, which has its own small but significant catchment. Other streams, such as the Tang Ball Beck, contribute water locally as well.

Rainfall in April and May 2000 reached exceptionally high levels. While the summer months were reasonably dry, autumn levels greatly exceeded long-term averages. Because of the spring rainfall, Yorkshire soils had failed to dry to average levels by late summer. A series of depressions resulted in several bursts of concentrated rainfall.

On 29 October, a warning was given of an expected rise of 3.1m above normal. The river rose quickly over a five-day period to a peak of 5.4m above normal on 3/4 November. There was a second peak of 5.1m on 8 November.

In the two weeks following 29 October, 353 properties were flooded (not including garden areas), 25 roads were closed, and many other areas were badly affected by the presence of the flood waters. Some outlying satellite villages were cut off.

The river defences at York included about 4km of flood defences constructed during 1985–95; of this, two-thirds were earth banks, while the remainder were constructed as concrete walls. There is also a barrier to prevent Ouse water from backing up the River Foss (Foss water is pumped over the top of the barrier at up to 30 tonnes of water per second). The river level assumed for design purposes was 5.45m above normal summer level. It is estimated that reinforcement of the flood defences immediately before the peak flood level saved about 3500 properties. At times, only the additional height of filled sandbags on top of the defences was holding back the water. Strategic control by the police force (triggered by a pre-agreed river rise to 4.6m), supported by the local authority, dealt with evacuation, relocation of pumps, and deployment of sandbags. Five old "Green Goddesses" and a fire tender from an air museum were called into play to augment normal pumping capacity. The Foss barrier pumps operated at full capacity, while engineers stood by in case of any failure. Army resources were also deployed.

While there was much disturbance and many properties were damaged, overall the city escaped lightly. Problems were caused by the failure of flap valves and other factors, and police divers, working in polluted water, played an important role in dealing with these.

Lessons were learned and, in 2004, the council published a comprehensive review.

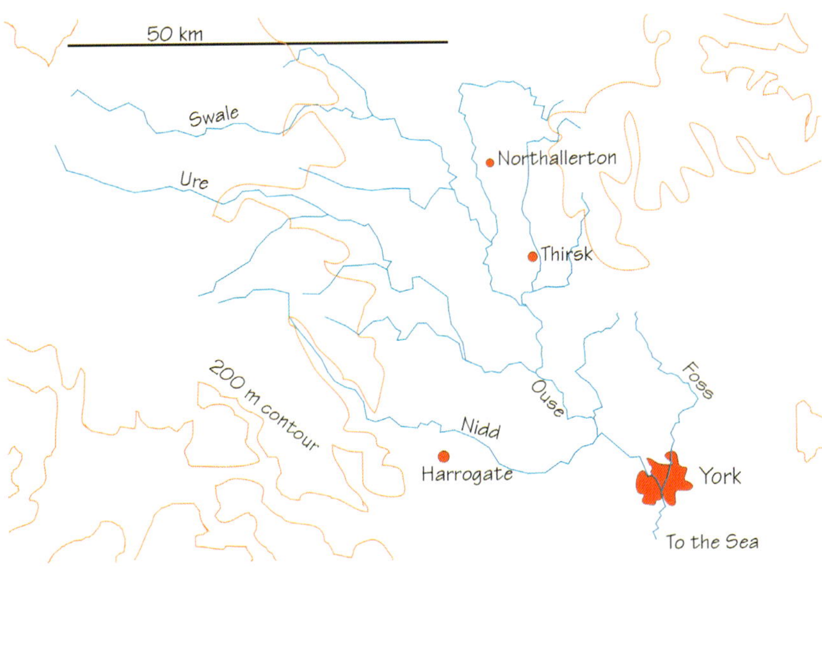

Figure 22: The Yorkshire Ouse drains a huge area (minor tributaries are not shown)

Figure 23: Damage at Boscastle (photo by www.apexnewspix.com)

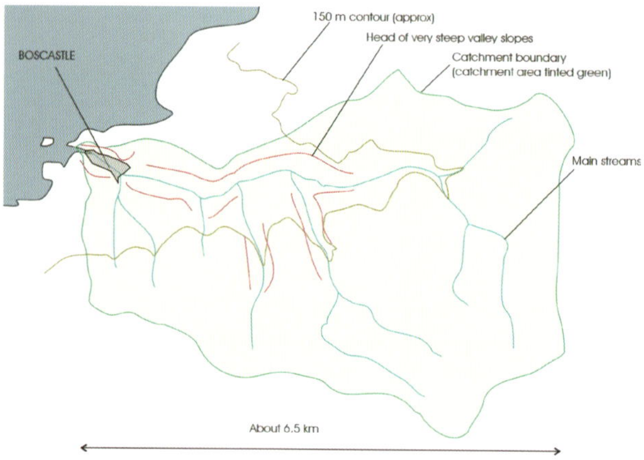

Figure 24: Boscastle river catchment

Boscastle — 2004 (and Lynmouth — 1952)

The following is based on my article *Danger on the Waterfront*, which appeared in Estates Gazette on 11 September 2004.

The situation in Boscastle, which received heavy media coverage (figure 23), greatly resembles that of Lynmouth, where disastrous flooding occurred in 1952 (coincidentally, on the same date, 16 August).

Both settlements included areas of development (including popular tourist facilities) that closely followed the courses of streams. In each case, the stream system that flooded drained an area of high land with relatively impervious subsoils. Also, in each case, the streams are steeply graded and occupy the bottom of narrow, steeply sloping valleys (figure 24). A further similarity in each case is that heavy and localised rain fell in a short period following earlier wet weather that had saturated the topsoil. This meant that most of the fresh rain immediately ran off into surface water channels. As a result, a very high proportion of the rain falling during the storms was channelled, with only minimal delay, through the settlements that straddled the streams near their mouths. The concentration of the water in narrow valley bottoms means that the devastation is limited in extent, but concentrated in effect.

True flash floods developed at both Boscastle and Lynmouth. Little or no warning was given and the rise in water level was sufficient to demolish buildings and carry vehicles downstream. During such flash floods, much of the erosion of the stream bed takes place and the "bed load" of pebbles and even quite large boulders is moved a significant distance downstream, together with tree trunks and other material. To quote a visitor to Boscastle, "There were 60ft trees just going down the river like matchsticks". As the current wanes, this bed load is dumped, so that a mass of debris is left. The larger items then remain in position until a sufficiently powerful flood moves them on again.

The assessment is that up to 100mm of rain fell within around 11 hours at Boscastle, about 77mm falling in about two hours. One source quotes 200.4mm as falling at Otterham, just to the east of the Boscastle stream catchment. In the case of Lynmouth, the estimate of rainfall was about 230mm in 24 hours. At Boscastle, a three-metre rise in water level (generally described in the media as a "wall of water" carried all before it, including vehicles (figure 23), buildings and shop stock. An early estimate of the losses at Boscastle was £50m (fortunately no life was apparently lost), while some 20 business properties were destroyed and the remainder of the tourist season lost.

At Lynmouth, the "wall of water" in the East and West Lyn rivers killed 34 people and destroyed numerous bridges, as well as sending 130 cars into the sea. Dozens of buildings were destroyed or irreparably damaged. A boulder of about 7.5 tonnes was found in the devastated basement of a hotel.

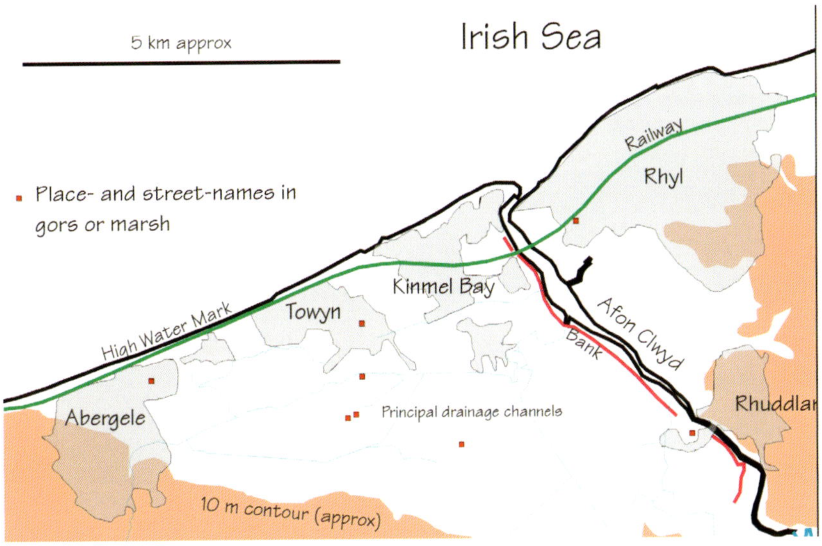

Figure 25: The Towyn area (most minor drainage channels have been omitted)

Without the benefit of a long river course with floodplains capable of absorbing some of the excess water, or a permeable subsoil also capable of absorbing water, the effect of such heavily concentrated rainfall can be catastrophic.

Arguably, the combination of circumstances necessary to cause such disasters is unlikely to be repeated within a foreseeable period. On the other hand, there is no logical reason why it should not. There are some general indicators that can be taken into account. For instance, the north Cornwall-West Devon area features generally high ground (about 150m) close to the sea, producing short, steep valleys that can rapidly channel run-off to sea level. Many localised floods in these valleys have failed to attract media attention, so that while insurers will be aware of them where damage has been caused, surveyors and valuers who are not immediately local may not be. In these and other areas where the topography produces flash-flood tendencies, and where increase in run-off is caused by hard surfacing or removal of vegetation, the risk is increasing. In upland Britain, a known general decrease in peat cover reduces the capacity of moorland to hold rainwater in storage. Side-effects of flash floods can include increased landslide risk, for instance where soft subsoil has been eroded away. The potential losses are great, and the possibility of insurance difficulties is increasing.

It is now incumbent upon surveyors and valuers to pay greater attention to the chances of occasional devastating floods in areas such as Boscastle, as well as warning of more obvious floodplain locations. Both rural and urban planning should take account of factors that may lead to increased run-off, such as downhill ploughing, scrub / woodland clearance and impermeable car parks. Such planning, combined with insurance pressures, should lead to some lessening of the risk.

Towyn — 1990

Towyn in Clwyd (figure 25) is a low-lying settlement next to the Irish Sea, and separated from it by the North Wales coastal railway. Until about 1800, the area comprised marshland, as suggested by the place names on the map and the presence of a closely spaced network of drainage channels. The shortage of level ground in North Wales tempted developers to build in this low-lying area, and many holiday and retirement homes were constructed during the 20th century. The existence of the sea barrier formed by the railway embankment, which had "always been there", led to an unthinking complacency.

On 26 February 1990, a deep depression (951 millibars) approached Britain. In the Irish Sea, very strong north-westerly winds were generated at a time when the highest tides of the year were expected. As high tide time approached, the railway embankment failed and seawater poured over the broken embankment, flooding about 10km^2 in Towyn and the adjoining settlements, including Kinmel Bay, to a maximum depth of about 2m. Nearly 3000 properties were flooded. The situation was compounded by the fact that many householders had no contents insurance.

The (then) Colwyn Borough Council, in its subsequent report on its involvement, recorded: "people become complacent over the years when nothing of note arises".

6 Assessing Flood Risk for a Site

A range of desktop and field risk assessment techniques are suggested, with a set of hypothetical examples.

Many sites will have no obvious risk except from local factors, such as blocked drains. In other cases, a more or less detailed investigation may be necessary. Some of this can be readily accomplished at your desk.

Client requirements

River bank and beach locations have long been sought after, for reasons of visual amenity or for access to boating or other recreation. For this reason, residential purchasers may choose to accept or even ignore the obvious risk of such locations. Commercial premises will also need to be in such locations for generally similar reasons (views from the hotel balconies, riverside dining, chandlery, tourist products and the like). In such cases, the risk must be a calculated part of the business plan.

Personal knowledge

Your personal knowledge of any particular area is probably the best starting point. More mature surveyors, in particular, will have memories of past flooding episodes, in some cases for 40 or 50 years into the past. Especially if they have been personally involved, such incidents may be graven deeply into the memory. However, the construction of flood defences or other redevelopment since previous floods may require reconsideration of any particular location.

The mature surveyor may also remember the topography of areas before they were developed. This may suggest areas where long-dry, cut-off river or stream channels (for instance) may be reoccupied by standing or flowing water at times of high water table or overbank flooding.

Research

The next phase of desktop study should probably begin with consideration of the maps of flood risk issued by the Environment Agency (see below). While these cannot be perfect because of the variables involved, they do provide an excellent starting point, certainly in respect of those areas that are not known in intimate detail to the surveyor.

The surveyor then needs to study topographical and geological maps. For immediate consideration, the Ordnance Survey (OS) 1:50,000 map can be used, as this will indicate the general locations of low-lying areas. However, it is a risky tool for assessing the level of hazards at any given point. For such purposes, the OS 1:25,000 map, or better a 1:10,000 map, should be used as this will indicate much finer topographical differences. Remember that a metre or so may be critical.

You may, of course, have access to sophisticated map databases or Geographical Information Systems, but these are expensive for small practices to operate. A collection of paper maps may suffice (keep the old ones, as they show old features that may prove to be important later, such as streams that become culverted).

On Geological Survey maps that show superficial deposits (do not make the mistake of buying a map that shows only "solid" or "bedrock" geology), alluvium is usually shown as a very pale yellow or cream tint. Different maps may have slightly different colouring schemes, so be careful to read the map key. As a good rule of thumb, the presence of alluvium (sediment deposited by streams or rivers) suggests the presence of a floodplain. Peat (usually a pale brown tint) in lowland areas suggests low-lying land with poor drainage. Cross-hatched areas (check the map key — different types of hatching indicate different features) may show made ground (landfill etc) that places a site above natural flood levels. In coastal areas, the map may show other features, such as storm beach deposits. You will need some practice to interpret some of these features meaningfully.

Areas shown on the maps as river terrace deposits (usually a pale orange-brown tint) are normally above current flood levels; they are remnants of earlier floodplains formed before the river cut down by erosion to its present course.

If you have not had much occasion to consider flood risk in the context of map study, it will be of little use to charge at it without consideration. You should start beforehand by making a detailed study of large-scale topographical maps and relating them, in the field, to the natural and artificial features that you can see. Those who are experienced walkers, and already accustomed to reading detailed maps, will have a distinct advantage. River cliff scars, railway embankments, constricting culverts, river defence walls and embankments, and other features must be related to the contours and other symbols on the map so that you can see instinctively, after practice, exactly where floodplains extend and how the surrounding topographical features, both natural and man-made, impinge upon them.

Take particular care to identify low rises that have historically provided adequate protection for long-established farmsteads, churches, village centres and the like. If they have always provided protection, these rises will almost certainly continue to provide protection in the absence of major change elsewhere (but remember to take into account the potential effects of sea-level rises in those areas that are more likely to suffer from marine flooding). Figure 26 shows a simplified map of part of a British floodplain. The floodplain is taken as synonymous with the extent of present-day alluvium (tinted). The distribution of old settlement cores (taken as those settlements marked by churches with spires or towers) shows that several villages are sited just off the floodplain, but none are on it. Our forebears clearly had considerable understanding of flood risk.

A mesh of closely spaced watercourses, especially if on a gridiron plan, is a sure sign of low-lying, poorly drained land in rural areas. Therefore, it is advisable to check the spot heights.

The symbol for marshland is naturally a warning sign.

Once you have studied the map, ask yourself to what extent flood defences depend upon active measures, such as pumping, or passive measures, such as walls and non-return flaps, which could be differently affected by adverse weather, strikes, terrorism or other influences.

Are there actual or proposed works that could alter the entire flood management scheme, for instance the Lavant diversion system at Chichester? Are there defences that are not immediately obvious, such as the Piddletrenthide "bypass" tunnel?

Also, take note of sea defences: are they actively maintained and do they protect substantial urban or commercial areas? If so, maintenance is likely to be continued. Can the sea defences be outflanked in other areas nearby? Are there any proposed developments (such as marinas or other maritime developments) that might affect the behaviour of the sea and the natural distribution of settlement?

In areas close to the coastline, spot heights may be provided in otherwise featureless areas. These can be of great value. As an axiom, you can take it that anything below five metres above Ordnance Datum is at risk. If you examine OS maps of low-lying areas,

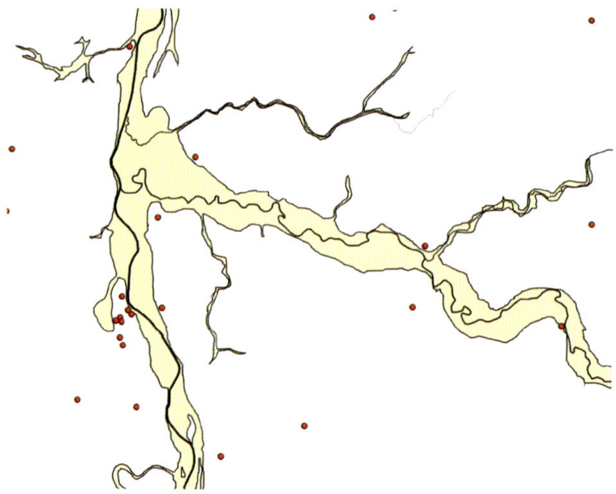

Figure 26: Floodplain churches

Figure 27: Blocked screens on culverts rapidly cause backup

you will often find areas of spot heights as low as one or two metres. Remember that these spot heights have been rounded, and that, especially well away from the immediate coastline, where seawater may be flowing with some velocity, half a metre or so could make all the difference between no damage, or only slight damage, and catastrophic flooding.

Develop your own collection or database of information relating to floods, the more detailed the better.

Remember that water level alone is not the only criterion; a property that is located at the extreme edge of a flood zone is likely to be only gently flooded, with most of the damage due to wetting and ruining of finishes and contents. A property, on the other hand, that is close to an existing channel may be completely destroyed by rapidly moving water, or even swept away.

In the field, you should develop an eye for the critical features that relate a given site or property to its surroundings; what seems like a level surface as you arrive in your car may, in fact, be a slope that is sufficient to put a property well above any likely flood level. Do not take the map for granted; seek personally for evidence of culverts (figure 27), small stream channels and other features that may not be apparent from the map. Also, the state of these features may give a clue to the local attitude to maintenance of drainage and other channels. After prolonged rain, the presence of standing water and saturated fields may suggest that any further rain could cause serious problems, while the opposite could be most favourable.

Some place-names are tell-tale signs, such as Marsh, Fen, Cors / Gors (in Wales — figure 25) or Carse (in Scotland). Do not ignore these warnings!

Unless you are working from an up-to-date topographical source, you may not be able to relate your position precisely to the map when in the field. In

Flooding from rivers or sea without defences
Extent of extreme flood
Flood defences
Areas benefiting from flood defences

Figure 28: A typical Environment Agency flood risk map and key

such cases, even a cheap hand-held GPS device should, unless you are hemmed in by tall buildings or thick woodland, locate your position to within 50m, or if you are lucky to within 6m or less.

Take time to assess the site in the context of the wider landscape. Is it potentially within the influence of a major river channel? If so, that may be the end of your research. Is it on chalk land where streams rarely, if ever, flow? Or is it in a clay vale where heavy rain soon results in saturation and overland flow? Is there heavy development in the area, with substantial hard surfaces? If there are steep slopes nearby, are they forested or bare? Are the river channels natural, with obvious washland, or are they managed?

Finally, do not forget to check basements. Properties that are satisfactorily dry at ground level, but have basements, may be affected by rising water tables (for instance in central London) if they are not tanked.

A sample Environment Agency map

Figure 28 shows a real Environment Agency flood risk map (reproduced with permission) and its key. The topography is partly obscured by the heavy tints used, although it will be known to a reader familiar with the area. Accordingly, for some flood maps you may need the appropriate OS map to hand. The extent of possible flooding is clearly marked, as are the river defences. However, the apparently sharp boundary of the risk area cannot be relied upon precisely, and you must report to your client accordingly. Note that to the left (west) of the image, there is a wide zone marked as being at risk only in extreme floods. To the right (east), there is only a narrow, discontinuous strip of such risk. This is because low river cliffs bound the floodplain for much of its eastern flank, while to the west the ground rises more gradually. In areas of such gentle slopes, the precise limit of flooding will naturally be less easy to define.

On this map, provision of river defences is extensive (the sinuous double trace of defences defines the otherwise rather obscure river course for much of the image). If you are unfamiliar in detail with the area, a visit to the defences may be in order.

A simple exercise

Figure 29 shows a hypothetical situation where a river is leaving an area of low hills and beginning to develop a floodplain. Assume the scale to be 1:25,000 and the contour vertical interval to be 5m. For simplicity, no other topographic information is shown. Notional properties are marked, lettered A to H.

- **Property A** is located next to a tributary stream that has cut a shallow valley. The slopes rise more or less regularly from the stream and there is no obvious floodplain. By interpolation, the property is about 2m above the water. The probability of significant flooding from the apparently small channel is fairly low.
- **Property B** lies about 10m above the main river, at a point where no significant floodplain appears to be present. The risk of flooding is low, but there may be potential problems with erosion of the banks, which should also be considered.
- **Property C** lies on a low cliff that appears to be undergoing active erosion by the river. The site lies some 25m above the river and flooding is out of the question. However, site stability should, again, be considered.
- **Property D** is in an area where there is clearly a floodplain but, due to the positions of the contours, there is no obvious edge to the floodplain. Closer investigation is required and it may not be possible to make a proper judgment without a site visit.
- **Property E** lies at the head of a low bluff, about 20m above the floodplain. The bluff appears to be an old river cliff, but the river channel is now distant. Flooding risk is probably non-existent and, provided the bluff is in itself stable, this site should be secure for the foreseeable future.

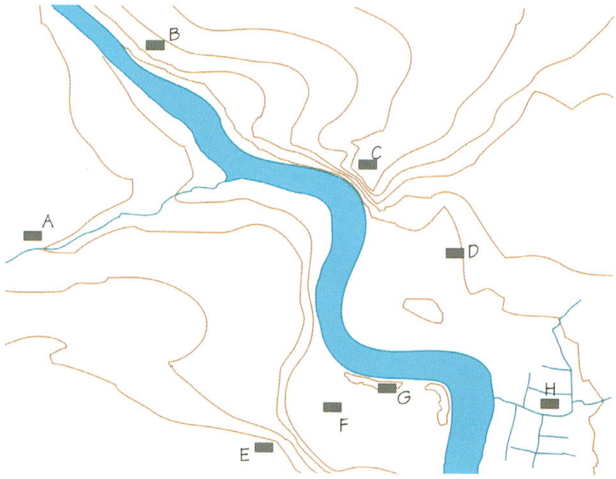

Figure 29: Reading the map

- **Property F** lies on the floodplain, the edge of which is clearly marked by contours that indicate a low bluff varying in height from about 5–20m. It is axiomatically at risk of flooding and you should consider the natural and artificial defences that may be present.
- **Property G** lies on what appears from the contours to be a natural levee. It should be safe for most of the time, but will be at risk in the most extreme floods. Also, the levee may be subject to active erosion, so that the site itself may not be stable within the scale of a human lifetime.
- **Property H** is in an area where the floodplain margin is not clearly defined by contours. However, being surrounded by a network of drainage ditches and almost certainly on the floodplain makes it at risk.

In practice, you may find it difficult to disentangle the necessary information from the other features on the map, particularly in built-up areas where the contours or smaller stream channels may be obscured by other detail and heights may be difficult to work out. Repeated study will make matters much easier to understand.

Reporting the risk

It is, of course, impossible to apply any meaningful numerical assessment of the risk in most cases. Clients may speak of statistical flood threats, but you should dissuade them from placing too much weight on quoted figures.

At the same time, try to relate the risk to everyday, easily understood perceptions of risk. For instance, you could speak of the risk of crossing the road. If you cross the road often enough, sooner or later you will at least receive a nasty fright. However, that does not stop most people from making the crossing. Everything involves risk.

Index

A

Afforestation 12
Alluvium 8, 26
Aquifers 5, 6
  Distribution and form 7
  Performance 8
  Recharge area 6
Aquicludes 5, 6

B

Bankfull 10
Barrier beach 14
Basements 8, 27
Bed load 22
Boreholes 6, 7, 8
Boscastle 11, 22–23
Bournes 7
Wellesbourne 7
Winterbournes 7

C

Catchment, river 11, 19, 20, 21, 22
Cave systems 6, 7
Chesil Beach 14
Clay 8, 19

D

Dams 6, 7, 17
Dendritic watercourses 5
Dunes 14

E

Environment Agency
  Online 1, 2, 12, 25
  Reports 16, 20

F

Fens 13, 15
Flash flood 11, 23
  Boscastle 11,22
  Lynmouth 22
Flood
  Control 12
  Definition 1
  Forecasting 12
  Management systems 11
  Perception of risk 12
  Significance 1
Floodplain, development of 8–11
  Deposits 9, 10
Flood, risk assessment 25–28
Flood, types 11
  Flash 11
  Global warming 11
  Groundwater 11
  Marine 13–16
  Sheet 11
  Unusual 17–18
Flood warnings 12
  Boscastle 22
  Lewes 20
  Lynmouth 22
  York 21
Flush 6

G

Geographical Information Systems 26
Geological Survey maps 26
Global warming 11, 12
Groundwater 6, 8
  Flooding 11

H

Hallsands 14
History 26
Hydrology 5
Hydrological cycle 5

I

Insurance 2, 12
Intra-urban flooding 1
Isostatic rebound 15

L

Landslides 17, 23
Levee 10, 11
  Artificial 10
Lewes 19–21
Lynmouth 22

M

Marine flooding 13–16
Meanders 8, 9, 10
Meteorological forecasting 12
Mine 7
Bursts 17
Motorway cuttings 7

N

Neap tides 13
Norwich Union (Aviva) 1
  Digital elevation model 2

O

Office of Science and Technology 1
Ordnance datum 26
Ordnance Survey maps 25, 26
Overbank 10
Oxbows 10

P

Peat 8, 9, 26
Piddletrenthide 12
Place names (indicative) 27

R

Railway embankments 7
  Lewes 21
  Towyn 23
Rainwater 5
Recharge area 6
RICS 2
Risk
  Perception of 12, 16
  Reporting 28
River catchment 11, 19, 20, 21, 22
River Darent 7
River Mole 7
Rocks
  Engineering properties 8
  Permeable 5, 6, 8
  Porous 5, 8
Royal Institution of Chartered Surveyors 2

S

Sea defences 14–15
  Artificial 14–15
  Natural 14
Sea level
  Changes 15–16
Mean 13
  Ordnance datum 26
Sediments
  Riverine 9, 10, 12
  Marine 14
  Shrinkage of 15
Severn Valley 12
Sheet floods 11
Spot heights 26
Spring 6
  Spring-line settlements 7
Spring tides 13
Storm surges 15
  1953 surge 15
  Thames Barrier 15
Streams, seasonal 7, 11
Subsidence 8
Superficial deposits 8
  Alluvium 8
  Glacial 8
  Peat 8, 9

T

Thames Barrier 15
Thames Gateway 2, 16
Tides 13, 15
  Neap 13
  Spring 13
Towyn 23
Tunnels 6
Tsunami 18

W

Washland 12
Watercourses 5
  Dendritic 5
Water cycle 5
Water flow 5, 6, 7, 8, 12
  Meters 12
Water residence times 8
Water table 6
  Behaviour 7
  High 6
  Low 7
Wellesbourne 7
Wells 6, 8
Winterbournes 7, 11, 20

Y

York 21